Kids & TECHNOLOGY
MISSION 21

DESIGN

SHARON A. BRUSIC

Developed by the Technology Education Program at Virginia Polytechnic Institute and State University.
John G. Wells, Research Assistant; William E. Dugger and James E. LaPorte, Co-Directors
The Mission 21 Project was funded by a grant from NASA.

DELMAR PUBLISHERS INC.®

Special Thanks

With special thanks to Robert H. LaJuett for his valuable assistance in preparing the manuscript, and to William E. Dugger, Jr. and James E. LaPorte for their roles in making this Mission 21 series possible.

Photography Credits:
p. 3: (top) Courtesy of The Little Tikes Company; p. 9: Courtesy U.S. Divers Co., Inc.; p. 10: (top) Courtesy Gallagher Power Fence, Inc., (bottom) Courtesy Walpole Woodworkers, Inc.; p. 12: (top) Courtesy Northern Homes Inc., Hudson Falls, NY, (bottom) Courtesy Creative Images by Harley/Harley Herold; p. 13: Courtesy Temple-Inland, Inc.; p. 14: (top) Courtesy Reynolds Metals Company; p. 16: Courtesy The Little Tikes Company; p. 18: (top) Courtesy PiMS; p. 19: Courtesy Nike, Inc.; p. 22: (top) Courtesy The Clorox Company, (bottom) Courtesy Ball Corporation; p. 23: Courtesy of AT & T Archives; p. 26: (middle right) Courtesy Creative Images by Harley/Harley Herold, (bottom upper right) Courtesy of Kellogg Company; p. 28: (top) Courtesy Eaton Corporation; p. 29: Courtesy Steve Sartori, Syracuse University Photo Center; p. 30: (top) Courtesy Ford Motor Company, (bottom) Courtesy Creative Images by Harley/Harley Herold; p. 31: Courtesy Alan Jefferis.

Cover photo by Doug Persons

Freelance Developmental Editor: Pamela Fuller

Delmar Staff
New Product
Acquisitions: Mark Huth
Senior Project Editor: Laura Gulotty
Production
Coordinator: Barbara A. Bullock
Senior Design
Supervisor: Susan C. Mathews
Photo Research and
Acquisitions: Michael Nelson
Art/Design Supervisor: Judi Orozco

For information, address Delmar Publishers Inc.
3 Columbia Circle, Box 15-015
Albany, New York 12212-5015

Printed in Mexico
Published simultaneously in Canada
By Nelson Canada
A division of The Thomson Corporation

1 2 3 4 5 6 7 8 9 10 XXX 99 98 97 96 95 94 93

ISBN: 0-8273-4950-5

CONTENTS

Did you know?
Hammers are designed to do two jobs. You pound nails with one end. You use the claw end to pull out nails.

PLANNING IDEAS

Look around at all the things you have and use. You have shoes and clothes. You use books, pencils, and toys. All of these things must be designed. Many other things are designed, too. Buildings, tools, and bicycles are some examples. Can you think of others?

This toy is designed to let kids build houses. What is the toy called? (Answer on page 32.)

People design things before they make them. **Designing** is planning out ideas.

People think about how things should look. They decide whether buildings should be tall or short. They decide the color of your pencil. They even choose the size of the door knob!

People decide how things should be made, too. Will we make plastic toys or wooden ones? Should the book have a hard cover or a soft one?

There are lots of things to think about when designing. You will explore many of them in this book.

DESIGN IT!

People who design things are called **designers.** Designers think about what people want and need. For example, a child may want something that will hold crayons. A designer can design a new holder.

To begin planning, designers think about what is needed. They think about different things that can hold crayons. They consider

Did you know?
Fred Morrison designed the first Frisbee®. He threw pie tins as a young boy. He thought it was fun. He designed a plastic toy to use in the same way.

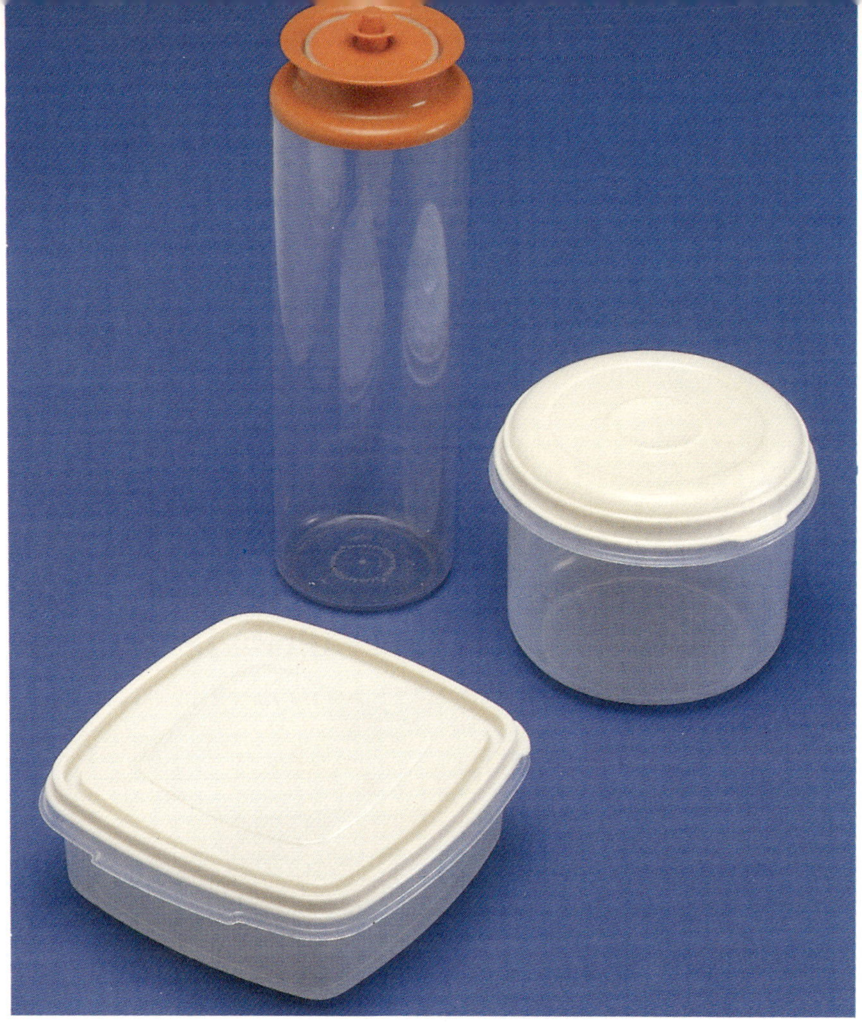

Look at these designs. Which one is best for holding uncooked spaghetti noodles? (Answer on page 32.)

many sizes, shapes, and colors. They draw lots of pictures and then choose the one best design.

Next, designers make a real example of their idea. This is called a **model.** A designer uses a model to test the design. Sometimes the designer sees a problem with the design. The design can be changed to make it work better.

You can be a designer, too. Think about something you need or want. Do you need a box to hold cards or toys? Do you want something to hold pennies? Think about it. Then, design it!

A GOOD DESIGN DOES THE JOB

Many things are designed to do a job. A can opener is designed to open a metal can. A bicycle is designed to move people on wheels. A key chain holds keys. A house is designed to protect you from weather.

Did you know?
Cartons like these are coated with wax or plastic. It keeps them from leaking.

▲ Scuba divers need special equipment to dive under water. Look at the things shown. Which one is designed to help the diver move fast under water? (Answer on page 32.)

Think about the stairs in your house. What are they designed to do? Staircases make it easy for you to move from one floor to another. Could you get to the second floor without them?

Stairs are designed so you can easily use them. Each stair goes up the same distance. That way you don't trip when you climb them. Most stairs in homes go up about 7 to 8 inches (17 to 20 centimeters) each. That's a good distance for most people. What if each stair was as tall as you are? Could you climb them?

▲

Did you know?
Cows get a small shock when they touch this fence. It keeps them from trying to get out. It's an electric fence.

DIFFERENT DESIGNS FOR DIFFERENT JOBS

Things are designed in many ways. All bridges do not look alike. They have different shapes. All cars do not look the same. There are many styles. All forks and spoons are not the same. They come in

different sizes and shapes. Some are metal and others are plastic.

Think about the fences you have seen. Some fences are tall. Some fences are short. Some are made of metal and others are made of wood. Some fences are just nice to look at. Some fences are used to keep animals inside an area. Describe the white fence in the picture.

One design will not fit every need. Sometimes many designs are needed. Different designs are needed for different jobs.

▲

This fence is very tall. It is designed so people cannot climb over it. Where is this fence most often used? (Answer on page 32.)

MAKE IT FIT

Designers must think about size and weight. Things are not useful if they are the wrong size or weight.

What if your pencil was as thick as your arm? Could you use it? What if your door

Did you know?
Most doors in houses built today are made the exact same height. They are 6 feet 8 inches (2 meters) tall.

Some packages are designed to hold a certain number of products. How many heads of lettuce fit into each box shown? (Answer on page 32.)

was only two feet tall? Could your teacher walk through it? What if your shoes weighed as much as you? Could you lift your feet?

You probably could not do any of these things. These objects would not be the right size or weight. They would not be useful this way.

Think about things you use. Imagine them in different sizes and weights. Picture tiny pillows and heavy bars of soap. Think about huge crayons and baseballs as light as feathers. It would be a strange world if things didn't fit just right!

THE RIGHT ONES

Things are made from many kinds of materials. Designers must choose materials that will be best for their product. How useful is a paper raincoat? Why are pots and pans made of metal? Imagine wearing wooden mittens!

Did you know? Most mail trucks are made of **aluminum** today. This type of metal does not rust. Mail trucks are designed to last about 24 years.

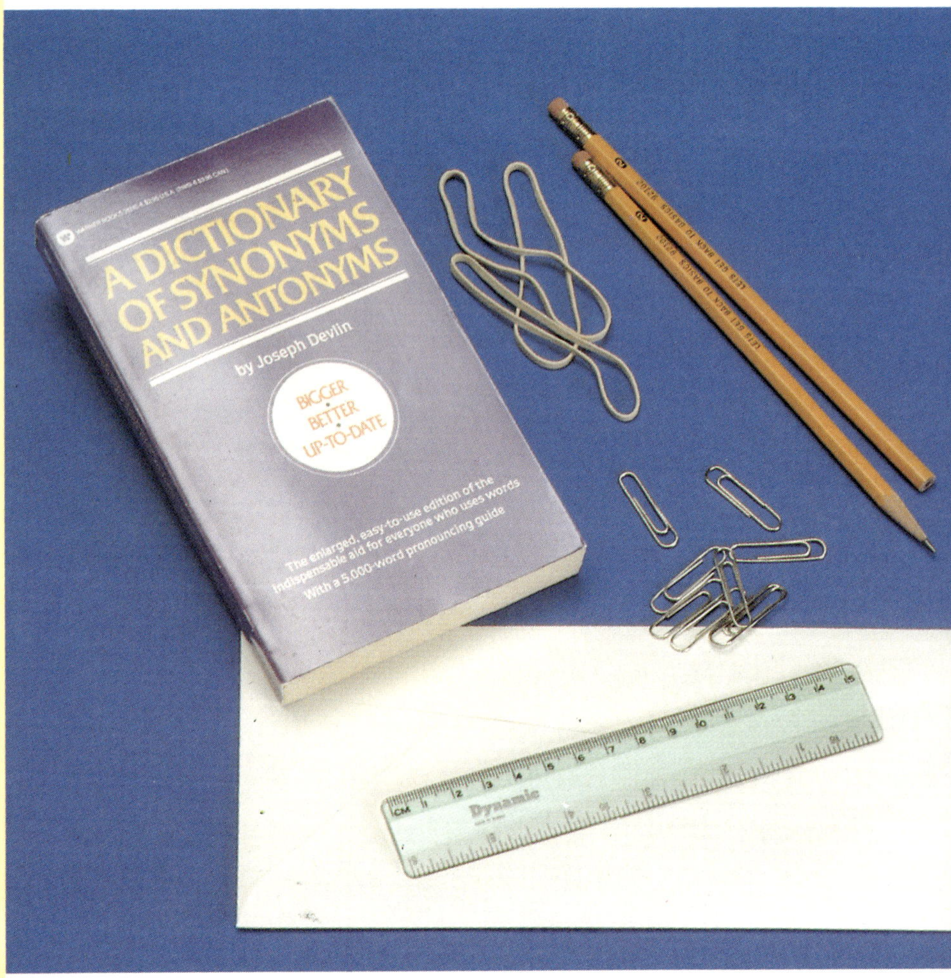

Notebooks like these are usually made of two materials. What are they? (Answer on page 32.)

Designers also choose materials that cost the right amount. Could you pay for a lunch box made of gold? Are wedding rings made of plastic?

Think about things you use every day. Some are made of plastic. Some are made of wood or metal. Other things are made of paper or rubber. See the photo to the left. Think about the materials used for these things.

There are many materials available. When designing, the hard part is picking the right ones for the job.

SAFETY FIRST

Have you ever heard someone say, "safety first"? People say that because they don't want you to get hurt. **Safety** is important. Designers must think about it, too.

Did you know?
Baby gates have safer designs today. Young children cannot get their heads caught inside the openings.

Why are most school buses painted this color? (Answer on page 32.)

Did you ever notice that small children's toys have smooth edges? Many toys are made of plastic, too. They do not break easily. Toys are designed that way so children don't get hurt.

Babies often put small objects in their mouths and ears. They can choke on small parts of toys or get them stuck in their ears. Designers know this. They design babies' toys so that there are no small parts. It is much safer.

All items should be designed with safety in mind. How are cars made safer? How many safety features can you find on a school bus?

Did you know? Many people want red sports cars. What color do you like?

LOOKING GOOD

People like to buy things that look good. They want good looking shoes. They want good looking cars. They even want good looking lawn mowers!

A good design looks appealing. Think about the color. Is it pleasing? Look at the

Is this a good COLOR?

Is this a good SHAPE?

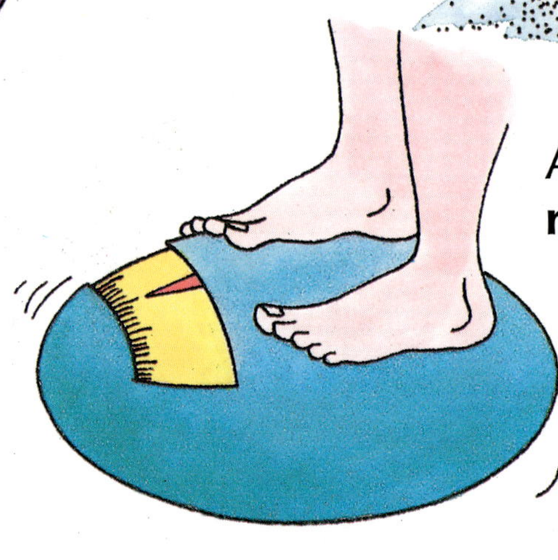

Are these holes the right SIZE?

A

B

C

shape. Is it fitting? Will it work well? Also think about the size. Is it right? Is it too bulky or too small?

Designers think about these things. They try to find the best looking design. They try to find the design that most people like. Sometimes they make many different designs. Then they can please more people.

Look at these shoes. Which one has the best design for walking through water? (Answer on page 32.)

LOOK AND LEARN

Let's take a look at a design. Look at the picture of the **umbrella.** Have you ever used one of these? Many people use them but not many people think about how they are designed.

Umbrellas come in many colors and styles but the design is much the same. They have similar shapes that make rain drip off easily.

Did you know?
A guitar body is hollow, not solid. This design is good. It makes the sound louder when the guitar is played.

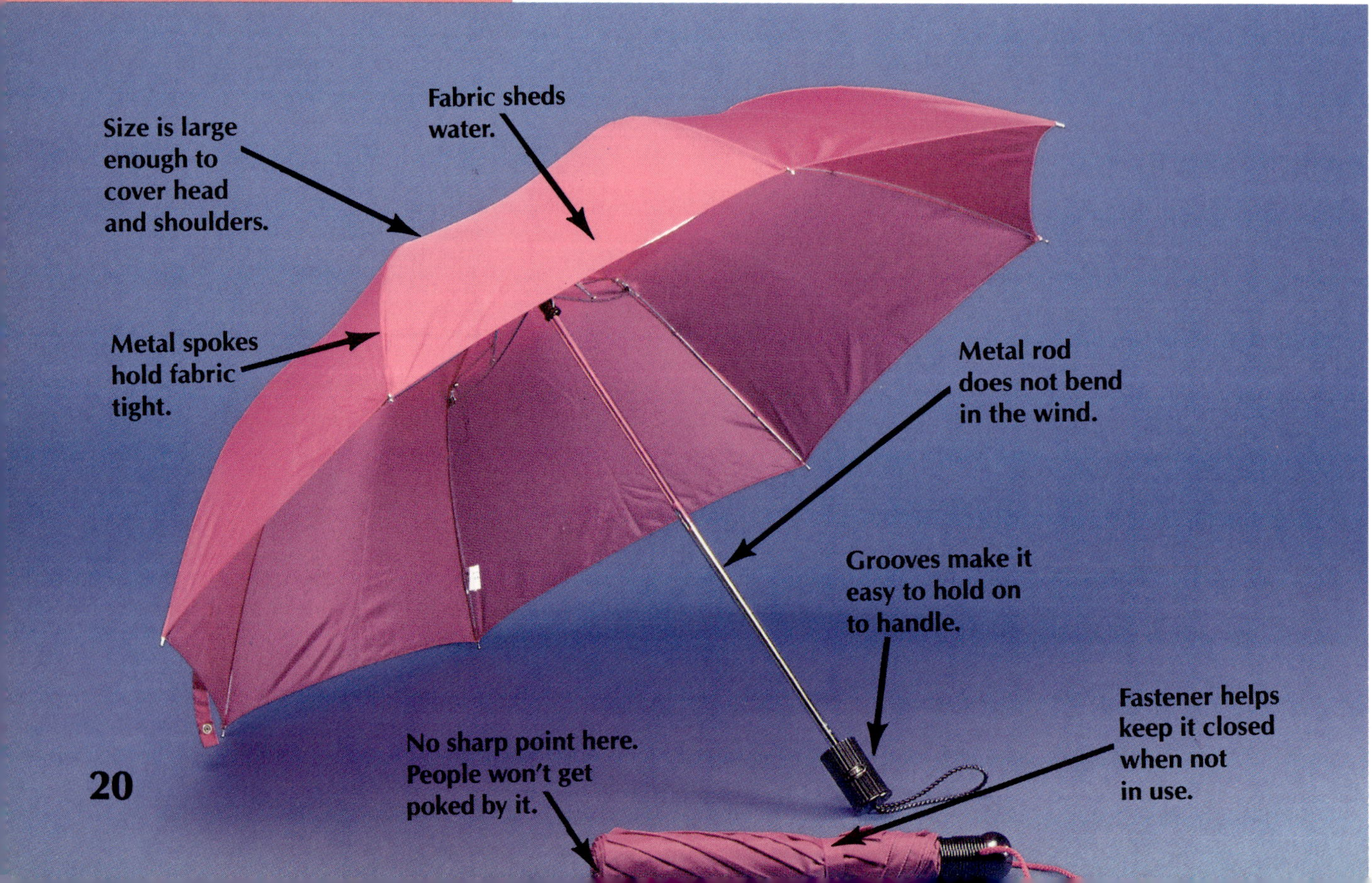

You can always see the gumballs in these machines. Clear glass or plastic is used. Why are they designed this way? (Answer on page 32.)

Read all the other notes on the picture. See if you learn something new about the umbrella design.

Designers often look at how other things are designed. They learn new things this way. You can learn a lot, too. Take a look at things you have and use. Study their shape and size. Check out the materials used. Look at how they work. Decide if it is a good design, and then think about ways to make it better.

▲ **Did you know?**
At one time, bleach was only sold in heavy stone or glass jugs. Today you can buy it in plastic jugs. They are lighter and they do not break easily.

OLD AND NEW

Designs are always changing. Designers try to find ways to improve things. Sometimes new designs solve a problem.

At one time, soda cans had a different kind of lid. People pulled on a metal ring to open the can. The metal piece came off. People often tossed these metal pieces on the

◀ This is an old design of a machine that you use today. People used this design many years ago. What is this machine? (Answer on page 32.)

ground. Other people put the metal pieces inside their soda cans. Then they drank their sodas. Some people swallowed the metal pieces!

You open soda cans in a different way today. You still lift the tab to open the can but the metal piece does not come off. It stays attached to the can. It is much safer!

Designers like to find better ways to make things. That's one way that we get improved products.

WOW! THAT LOOKS GREAT!

"Wow!" Billy said, as he looked through a new toy catalog. "That game looks great!"

Billy's parents bought the game for his birthday. But Billy was not too happy when he started playing with it. He found that the game was not as much fun as the picture made it look.

Did you know?
A good design is often simple. It lasts a long time. This soup label has hardly changed in almost 100 years!

This sign is designed to send a quick message. What's the message? (Answer on page 32.)

The designers of the catalog Billy saw were clever. They knew that pictures helped to sell toys. They arranged the picture so that the game looked super exciting. Billy fell for it!

Designing pictures, posters, and signs is no small job. Designers spend many hours finding the best way to set them up. They pay close attention to the words and colors used.

Look around at all the things that have words and pictures on them. What kinds of things capture your attention? What designs are exciting to you?

WRAP IT, BOX IT, LABEL IT

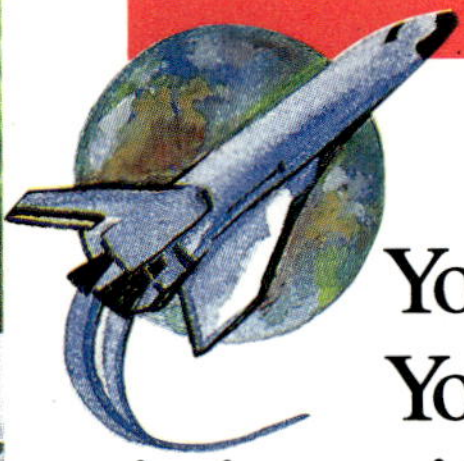

Your candy bars come in wrappers. You buy cereal in boxes. Pickles come in jars with labels on them. All of these things are forms of **packaging.** Did you know that packages are designed, too?

Did you know?
You get straws and forks in wrappers at some restaurants. They are wrapped to make sure you get clean ones every time.

Why is this new radio packaged this way? (Answer on page 32.)

Packages are designed to suit the things that go into them. Think about packaging eggs. Could you carry them home from the store in a shoe box? Why isn't ice cream sold by the bottle? It would be hard to scoop it out!

Labels are designed, too. They tell you about the things inside the packages. What if soup cans did not have labels?

You need to think about many things when you design packages. Will it keep the product safe or fresh? Does the package look good? Is the package easy to use?

OVER YOUR HEAD AND UNDER YOUR FEET

What's over your head when you are in school? It's a ceiling and a roof. A building is all around you. It gives you shelter. What's under your feet when you cross the street? It is a roadway.

Did you know?
Bridges need to be very strong. They must carry their own weight plus the weight of cars and people that cross them.

This building is called the Carrier Dome. It was designed for a specific use. What takes place in this building? (Answer on page 32.)

Buildings and roads are designed. Bridges are designed, too. They are designed to be sturdy and strong. They must last many, many years. They must be made safe, too.

These designers must think about many things. They think about wind and rain. Will it fall apart in a storm? They think about the weight of snow. Will the roof hold all the snow that falls on it?

Designing buildings and roads takes a lot of careful planning. Are you good at planning? Would you like to design buildings and roads one day?

WHEN I GROW UP, I WANT TO DESIGN...

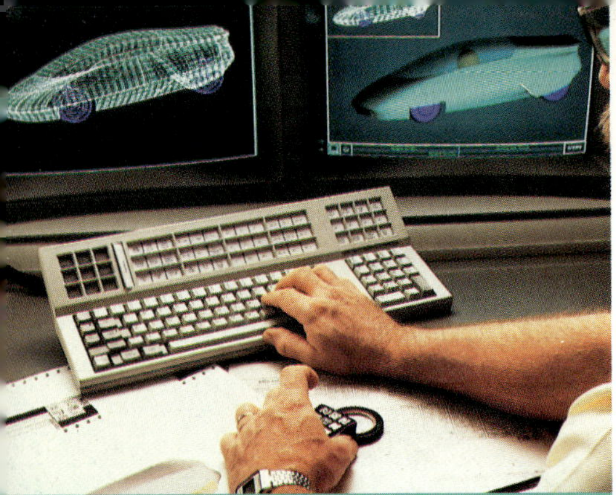

Finish the sentence in the title. What do you want to design? Do you have an idea for a new car or fun toy? You could design colorful books or cartoons. How about designing a new tool or a super tall building? So many things are designed. You might enjoy designing one of them!

Did you know?
Designers often use computers. A person is designing a new car. You probably ride in cars that were designed on computers.

This woman is an **architect.** Which of these things does she design—books, houses, or shoes? (Answer on page 32.)

It is never too early to start being a designer. Do you enjoy coming up with new ideas? Do you like to build things and be creative? Do you like to solve problems? You might just have what it takes to be a great designer!

Many jobs deal with design. Talk to adults you know. Ask them if they design things. Find out what kinds of things are designed in your town. You may be surprised to find so many!

ANSWERS TO PHOTO QUESTIONS

Unit Title	Question Page Number	Answer
Planning Ideas	5	Lincoln Logs
Design It!	7	The tall one
A Good Design Does the Job	9	Fins
Different Designs for Different Jobs	11	A prison or jail
Make It Fit	13	24 heads of lettuce (2 dozen)
The Right Ones	15	Paper and metal
Safety First	17	Bright yellow is easy to see by other drivers.
Looking Good	19	C
Look and Learn	21	People like to see the colorful gumballs.
Old and New	23	Telephone
Wow! That Looks Great!	25	No pets allowed
Wrap It, Box It, Label It	27	To keep it from breaking or getting scratched
Over Your Head and Under Your Feet	29	Sporting events and popular music concerts
When I Grow Up, I Want to Design...	31	Houses